How We Learn and Why We Don't

Student Survival Guide Using the Cognitive Profile Inventory

Dr. Lois Breur Krause
Clemson University

CENGAGE
Learning™

Australia • Brazil • Japan • Korea • Mexico • Singapore • Spain • United Kingdom • United States

CENGAGE Learning™

How We Learn and Why We Don't: Student Survival Guide Using the Cognitive Profile Inventory

Dr. Lois Breur Krause

Executive Editors:
Michele Baird

Maureen Staudt

Michael Stranz

Project Development Manager:
Linda deStefano

Senior Marketing Coordinators:
Sara Mecurio

Lindsay Shapiro

Production/Manufacturing Manager:
Donna M. Brown

PreMedia Services Supervisor:
Rebecca A. Walker

Rights & Permissions Specialist:
Kalina Hintz

Cover Image:
Getty Images*

* Unless otherwise noted, all cover images used by Custom Solutions, a part of Cengage Learning, have been supplied courtesy of Getty Images with the exception of the Earthview cover image, which has been supplied by the National Aeronautics and Space Administration (NASA).

For product information and technology assistance, contact us at **Cengage Learning Customer & Sales Support, 1-800-354-9706**

For permission to use material from this text or product, submit all requests online at **cengage.com/permissions** Further permissions questions can be emailed to **permissionrequest@cengage.com**

ISBN-13: 978-0-7593-2734-4

ISBN-10: 0-7593-2734-3

Cengage Learning

5191 Natorp Boulevard
Mason, Ohio 45040
USA

Cengage Learning is a leading provider of customized learning solutions with office locations around the globe, including Singapore, the United Kingdom, Australia, Mexico, Brazil, and Japan. Locate your local office at: **international.cengage.com/region**

Cengage Learning products are represented in Canada by Nelson Education, Ltd.

For your lifelong learning solutions, visit **custom.cengage.com**

Visit our corporate website at **cengage.com**

Printed in the United States of America

Table of Contents

Introduction to the student edition

So welcome back to school, or welcome to college, or glad to see you want to do something about your grades... or whatever fits your situation. In any case, the goal of this book is to help you to understand how you think, how you learn, and even, perhaps, why you don't.

Your first task is to recognize that everyone thinks a little bit differently from everyone else; in some cases, a lot differently. But you probably know that. In your years as a student you may have wondered why your instructor didn't say it clearly in the first place, when he or she finally got around to the concept. Or wished he wouldn't reprimand you when you made the smallest comment to the student next to you; after all, you were talking about the work. Sometimes you might have been glared at for doodling when the teacher just didn't understand that you think better when your pencil is moving. And why do they have to get so upset when you listen to your CD's when doing class work? What is it about gum that bothers them, anyway? And you really were awake, you just think better with your eyes closed. Why do they insist on your doing all that boring repetitious drill, or why don't they give you more practice on the type of problems they ask on the test?

All these questions are valid, and relate to how you like to or need to handle information you are trying to understand. We are going to explore the cognitive profile model for differences in how people learn. After you do the inventory and have drawn your own profile, we will explain what the letters and numbers mean and what they can tell you about effective ways to study for your profile. The most important thing to keep in mind as you do the inventory is that there are no wrong answers. No answer is better than any other answer. Just evaluate the words in the inventory by how you like to relate to the world, how you focus on something you want to learn, what you are interested in, and life in general. The more thoughtful and honest you are about your choices, the more valuable the information will be to you. Above all, don't make choices based on what you think you should be like, or what someone else told you that you should do, or even by what you've been taught before about what's good and what's bad. This is about you, and you are unique, and we need to know how you think, not how someone else might have told you that you should think.

At the end of the book we will spend some time on other choices that you will be making, such as choices of major and career path, based on your cognitive profile. But the first thing for you to do is take the inventory. It usually takes about 20 minutes, and then we'll go from there.

Cognitive
Profile Inventory

By

Dr. Lois Breur Krause

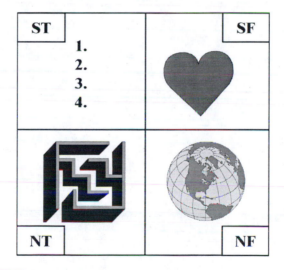

Chapter 1: Determining your Cognitive Profile

The Cognitive Profile Inventory is designed to help you to identify your personal style of thinking, learning, and making decisions. Although an individual's profile is subject to some change over time due to lifestyle, education, and other significant influences, the general shape of the profile probably won't change a great deal. But before too many details are discussed, the first step is to do the inventory for yourself. Once you have your own profile in front of you, the description of what it means will be much more meaningful.

The inventory consists of 60 pairs of words. You are to look at each pair, and choose which appeals to you more, or describes you better. As you are doing the inventory, keep in mind how you prefer to deal with life, learning, and people. It is important that your answers reflect how you really prefer to do things, not what you think you should do, or what someone tells you that you *should* do, or how someone else does things, no matter how much you admire them. No answer is better than any other answer. The best answer is the one that is right for you. This is your inventory, and it will be your profile. The more thoughtful and honest you are in self-evaluation, the more helpful the resulting profile will be for you.

For each pair of words, choose one, and give it a value between one and four. If you have a strong preference, you can assign it a higher value. You might think in terms of the following guidelines for values:

1. I generally prefer this some of the time, probably a little more often than the other choice.

2. I definitely prefer this over the other choice most of the time.

3. I strongly prefer this one, would very rarely choose the other

4. Absolutely! Would probably never (or almost never) choose the other if this choice were available.

For example, in flavors of ice cream:

Vanilla 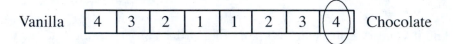 Chocolate

I personally would choose chocolate and give it 4 points, since I would almost always choose chocolate if my choice were between vanilla and chocolate.

If I liked vanilla a bit more than chocolate, my choice might look like this:

Vanilla 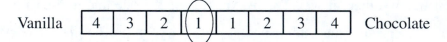 Chocolate

5. Circle one number in each row, to indicate your preference. Give points to only one word in each pair of words.

6. When you have finished each set of 10 pairs, add up the values you have given to the words in each column, and write the subtotals in the spaces provided with each set.

Step One: Taking the Inventory

Practical	4	3	2	1	1	2	3	4	Emotional
Facts	4	3	2	1	1	2	3	4	Feelings
Doing	4	3	2	1	1	2	3	4	Talking
Concrete	4	3	2	1	1	2	3	4	Personal
Read A Book	4	3	2	1	1	2	3	4	Tell a story
Get It Done	4	3	2	1	1	2	3	4	Enjoy doing
Roles	4	3	2	1	1	2	3	4	Loyalties
Sensible	4	3	2	1	1	2	3	4	Emotional
Protect yourself	4	3	2	1	1	2	3	4	Do for Others
Practice	4	3	2	1	1	2	3	4	Discuss
Subtotals: "A"=									="B"

Trial & error	4	3	2	1	1	2	3	4	Strategy
Protect yourself	4	3	2	1	1	2	3	4	Tell the Truth
Specifics	4	3	2	1	1	2	3	4	Overview
Concrete	4	3	2	1	1	2	3	4	Abstract
Read A Book	4	3	2	1	1	2	3	4	Work a Puzzle
Get It Done	4	3	2	1	1	2	3	4	Plan it out
Roles	4	3	2	1	1	2	3	4	Laws
Sensible	4	3	2	1	1	2	3	4	Logical
Information	4	3	2	1	1	2	3	4	Concepts
Practice	4	3	2	1	1	2	3	4	Understand
Subtotals: "A"=									="C"

Trial & error	4	3	2	1	1	2	3	4	Visualize
Realistic	4	3	2	1	1	2	3	4	Idealistic
Details	4	3	2	1	1	2	3	4	Big Picture
Touch, hold	4	3	2	1	1	2	3	4	Mental picture
Read A Book	4	3	2	1	1	2	3	4	Daydream
Get It Done	4	3	2	1	1	2	3	4	Create
Roles	4	3	2	1	1	2	3	4	Principles
Sensible	4	3	2	1	1	2	3	4	Logical
Protect yourself	4	3	2	1	1	2	3	4	Save the Earth
Practice	4	3	2	1	1	2	3	4	Think About
Subtotals: "A"=									="D"

Create	4	3	2	1	1	2	3	4	Share
Ideals	4	3	2	1	1	2	3	4	Relationships
Imagination	4	3	2	1	1	2	3	4	People
Possibilities	4	3	2	1	1	2	3	4	Loyalties
Listen to Music	4	3	2	1	1	2	3	4	Tell a story
Daydream	4	3	2	1	1	2	3	4	Group activities
Principles	4	3	2	1	1	2	3	4	Loyalties
Insights	4	3	2	1	1	2	3	4	Emotions
Save the Earth	4	3	2	1	1	2	3	4	Do for Others
Think about	4	3	2	1	1	2	3	4	Discuss
Subtotals: "D"=									="B"

Sharing	4	3	2	1	1	2	3	4	Strategy
Do for Others	4	3	2	1	1	2	3	4	Tell the Truth
Details	4	3	2	1	1	2	3	4	Overview
Concrete	4	3	2	1	1	2	3	4	Abstract
Tell a story	4	3	2	1	1	2	3	4	Work a Puzzle
Enjoy doing	4	3	2	1	1	2	3	4	Plan it well
Loyalty	4	3	2	1	1	2	3	4	Law
Emotion	4	3	2	1	1	2	3	4	Logic
Join a group	4	3	2	1	1	2	3	4	Lead a group
Discuss	4	3	2	1	1	2	3	4	Understand
Subtotals: "B"=									="C"

Strategy	4	3	2	1	1	2	3	4	Visualize
Theoretical	4	3	2	1	1	2	3	4	Idealistic
Experiment	4	3	2	1	1	2	3	4	Invent
Think	4	3	2	1	1	2	3	4	Imagine
Solve a Puzzle	4	3	2	1	1	2	3	4	Daydream
Plan it out	4	3	2	1	1	2	3	4	Create
Laws	4	3	2	1	1	2	3	4	Principles
Logic	4	3	2	1	1	2	3	4	Metaphor
Tell the Truth	4	3	2	1	1	2	3	4	Save the Earth
Analyze	4	3	2	1	1	2	3	4	Discover
Subtotals: "C"=									="D"

Step Two: Totaling your scores

Now write your subtotals for each letter "A" through "D" on the blanks below, and find the total for each .

"A"   "B"   "C"   "D"

_____   _____   _____   _____

_____   _____   _____   _____

_____   _____   _____   _____

Totals:

"A"= ST   "B"= SF   "C"= NT   "D"= NF

_____   _____   _____   _____

These four totals are your Cognitive Profile Quadrant Numbers, which describe how you learn, process information and make decisions.

Step Three: Plotting Your Numbers and Determining Your Profile

In Step Three, you will use the diagram below to plot your numbers in each quadrant and show your profile.

First determine which of your totals for ST, SF, NT or NF is your highest. *This is your "dominant quadrant."* Using the center of the diagram as zero, assign a value equal to your highest Quadrant Number to the farthest corner of each quadrant. For your dominant quadrant, your highest Quadrant Number, put a dot in the outermost corner of the quadrant labeled with that type. For your next highest number, find its position in the proper quadrant by estimating its proportion to your highest Quadrant Number.

Example: For a student with the following Quadrant Numbers, ST=45, SF=60, NT=32, NF=10: 1) The SF is the dominant quadrant, so the upper right corner has a dot, and it represents 60. Now each quadrant is zero in the center to 60 at the far corner. 2) The ST quadrant number is 45, and 45/60 is. 3/4, so the student would put the dot for ST three quarters of the way out towards the corner in the upper left quadrant of the diagram. 3) The NT quadrant number of 32 is a little more than half of 60, so the dot in the lower left quadrant would be just a little more than half way out from the corner. 4) The NF quadrant number of 10 is only one sixth of the highest quadrant number of 60, so the dot in the lower right quadrant is close to the center, only one-sixth of the way out.

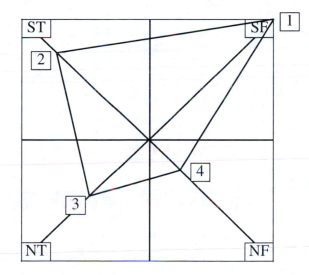

When you have plotted your points in each quadrant and placed your dots by estimating, connect the dots to make a quadrilateral figure.

Step Three

What's your style?

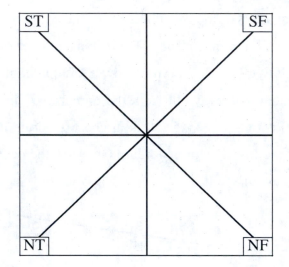

Use this diagram to plot your Cognitive Profile as in Step Three of the instructions. If you would like to color in your profile, color your area in the SF red, in the ST yellow, in the NT blue, and in the NF green.

NOTE: The inventory is also available on the web at www.cognitiveprofile.com. On the web, the calculations are done automatically for you, and you can print out your profile in color.

The Cognitive Profile Model

Dr. Carl Jung (1875-1961) was a Swiss psychologist who described personality by categorization. His original work is the basis for the widely used Myers-Briggs Type Inventory and much of the current research in individual differences in how we relate to the world around us. The Cognitive Profile Model that we use to describe differences in how we learn is based on Dr. Jung's personality type theories, with one major difference. Dr. Jung used four "bipolar descriptors," Introvert/Extrovert, Sensor/iNtuitive, Feeler/Thinker, and Judger/Perceiver. He believed that an individual's personality could accurately be described as a combination of one from each pair. For example, a person might be categorized as an ESFP, or Extroverted Sensor Feeler Perceiver. The Myers Briggs Type Inventory is a test that uses the same Jungian terminology and theory, and has been widely used in personality research. It categorizes each individual into one of 16 possible combinations.

But you know that sometimes, with some people or in some situations you will feel shy and introverted, and other times, with people you know or in comfortable surroundings you will be more outgoing. It's not always accurate to say that you are either one or the other. Most researchers who use Jungian models have classified people into one type and then said, that's it, that's what you are. The Cognitive Profile

Model recognizes that sometimes we live in one type and sometimes in another depending on the situation and what we have to do. We all have some of each type in us, just to varying amounts. Jungian types are also more complex than the Cognitive Profile Model types. Because they use four bipolars, the result is 16 possible combinations. You might be INFP or ISTJ, but you are considered to be always the same. That classification doesn't differentiate between the person who might always be extremely shy, and the person who is just a little shy around strangers in the beginning. However, according to other interpretations of Jung, both would be classified as introverted, although they are very different.

The Cognitive Profile Model uses only two of Jung's bipolar descriptors, the Sensor-iNtuitive, and the Thinker-Feeler. These two bipolars are drawn as the x (the Thinker-Feeler) and y (the Sensor-iNtuitive) axes on a grid. The quadrants or spaces are then labeled by the bipolars they lie between, and are called Sensor Thinker (ST), Sensor Feeler (SF), iNtuitive Thinker (NT), and iNtuitive Feeler (NF). (The N is capitalized in iNtuitive rather than the I to avoid confusion with Jung's Introvert.)

You took an inventory of 6 sets of 10 pairs of words per set, giving a preference score to one word of each pair. Then you plotted your numbers on the grid as one point in each of the four quadrants. Connect the points and you have a profile, a quadrilateral shape that shows some amount of area in each quadrant. Just by looking at your Cognitive Profile you can see that you have some area in each quadrant.

You probably have a dominant quadrant, or one quadrant in which you have the most area (your highest number). You can see that you have some area / preference / ability in each type of thinking, and you can strengthen those abilities by learning skills in each of those areas. This inventory does not attempt to determine or describe ability, but only preferences for certain types of activities. As we describe the quadrants and talk about you as one type or another, remember that sometimes you function in one quadrant and sometimes in another. You may be working in an ST way when you are doing math home-work, or organizing your notes or the membership list for your frater-nity, and living in your NF quadrant when designing the homecoming float or doing any other creative activity. Your dominant quadrant shows which type of activities you enjoy doing most, and how you pre-fer to tackle a task.

Chapter Two: Describing Your Cognitive Profile

The Styles

As we've discussed, the two bipolars we use in the Cognitive Profile Model are the Sensor / iNtuitive, and the Thinker / Feeler. The Sensor / iNtuitive is the y axis, or vertical line, in the diagram, and it describes how you think and how you handle information and ideas. The top of the y axis or Sensor, describes a person who is more concrete about things. If you are a Sensor, you get the idea of something new more easily if you can see, touch, smell, taste or hear it. You like to be able to pick a thing up and turn it around to see what's on the other side, how heavy it is, or what it feels like. At the bottom of the y axis is the iNtuitive learner. If you are iNtuitive, you probably make mental pictures of what someone is describing or what you are reading. You also probably think in pictures rather than in words most of the time and likely dream in color. You can picture something in your head, and probably "see" it turn around or move as it works.

The x axis, or horizontal line, represents the Thinker / Feeler. It describes how you make decisions. If you are a Thinker, you usually make decisions based on facts, details, and information. If you are a Feeler, you are more likely to make decisions based on emotions, personal values, and relationships. I usually describe this using a story. You're in your dorm room at 10 o'clock on a Thursday night, when some of your friends knock on your door. "We're going to Tiger Town

Tavern. Wanna come?" Before you answer, and in making that decision, you will generally ask yourself a few questions. If you are a Thinker, you will ask yourself things like "How much money do I have and what will it cost me? Will I have enough left over for what I need tomorrow? How late will I probably be out? Do I have everything I need prepared for tomorrow's classes? How early is my first class tomorrow, and will I be back in time to get enough sleep to make it to class and function as well as I need to? " If you are a Feeler, you would ask yourself questions also, but they would be very different questions. A Feeler would ask "Who is going to be there? Will I see people I want to see? Will I have fun? What am I wearing, is it OK or do I need to change? Will I make the right impression if I see people I want to see? What's the weather like? Will I get wet or cold or be uncomfortable?" Now, we know that they will both make the same decision, but Thinkers and Feelers have very different bases for making the decision.

An important thing to remember is that every individual has some area in each quadrant, which makes a *profile*. You are not limited to only one type of task or study technique. The Cognitive Profile model emphasizes that you have some area in each quadrant; some ability or preference in each area. You can draw on those areas you use less often as you need them and you can develop skills in those weaker quadrants. Everyone has times when you are surrounded by good friends and very SF, or working out a puzzle or problem in an NT way. You may make ST lists for some types of tasks, and you may be very creative in another area, such as music or art. No one is only one quadrant, but we tend to be stronger and more comfortable in one or two than in the others. It is this strongest quadrant where we will develop techniques to make our study hours more productive.

I should point out that every shape of profile is "normal." A perfect square generally means you haven't figured out yet what you are best at, and you are comfortable doing just about any kind of task. Those who have a perfect square when young often will develop a dominant quadrant as they go along in life and find their career paths and passions. In making choices of major and career, consider your talents, and think about what an 8 hour day on the job would involve, and is that something you would like to do?

A kite shaped profile, as in ST subdominant and NF dominant, or NT dominant and SF subdominant combinations generally means you have developed strong coping skills and realized they work for you (as organization in the ST case) or that you have very strong, very important friendships (in the SF case). In both these, the NT or NF is probably your "natural" dominant.

In the next section we will describe each quadrant, and you will, no doubt, recognize yourself in your strongest area. Think about what things you do that identify yourself with one quadrant or another. You will also likely find that those things described in your weakest quadrant are things you don't enjoy doing.

Sensor Thinker

If your dominant quadrant, (your highest number), falls in the ST quadrant, you are a sensor thinker. The **Sensor Thinker** (ST) is an analytical learner who works in an organized fashion, methodically and stepwise. You probably learn best alone, and by repetitious drill and practice. You have a profound need for timely feedback. Answers are either right or wrong, and discovery learning may easily frustrate you where there is no clearly defined path to the correct result. You probably memorize easily and well, and do best in recall tests. Because of their gift for retention of large quantities of information ST's may be expected to do well in law school or medical school. They can do very complex tasks exactly the same way each time, as in heart surgery, or landing a 767 aircraft. Think about it. If you were on the table having open-heart surgery, you would want the surgeon to do the procedure exactly the same way he'd done it successfully many times before. If you were piloting a huge commercial airliner with responsibility for hundreds of people on board, you would want to remember exactly how to land the plane without leaving out any critical steps in the procedure. Other ST's may enjoy business, accounting, or other professions that require attention to detail.

You may be happy to know, or may have already figured out that our schools are mostly designed for ST learners. In fact, the study tips that instructors or guidance counselors typically offer complement the ST's learning style. ST learners usually study best alone, in a well lit structured area, at a desk or table with a straight chair, with no distractions, and do repeated example problems and exercises. Please remember, when learning a concept that is presented in a way that doesn't lend itself to recall or memorization, you should restructure it. Break each concept into steps or small pieces, number each step, and be sure you master each piece before going on to the next. It may be helpful to copy large quantities of information onto flash cards.

Sensor Feeler

If your highest number was SF, you are a sensor feeler. The **Sensor Feeler** (SF) is also a concrete learner, and you are the type of learner for whom cooperative learning is ideally suited. You need to process information verbally, and learn best if you can relate personally to the content. When you need to process complex content, you should talk it through, and study with a friend whenever possible. However, after your session with a friend, you need to practice alone to be sure you know it yourself and can do it alone. You should break large tasks into several small steps, and find a means of relating to the content on a personal level. You might also create a story line or example problem using friends' and family members' names, or take part in a play or game. SF's often have a talent for relating to people, and find themselves in nurturing professions such as teaching, nursing, or social work. SF's tend to be nurturers, and do exceptionally well in caring professions such as teaching, nursing and other medical fields, social work, and anywhere that requires a depth of understanding and empathy for other people. However, some of these fields require a great deal of ST work in preparation which may be frustrating for the SF. You will need to develop those coping skills to succeed there.

Both SFs and STs need to learn sequentially, building from the known to the unknown with hands-on activities where possible. Teaching methods or study techniques that let you build new learning on a basis of previously well understood material work well for both ST's and SF's, in that the learning is built stepwise from basics to concept, and that the students build their own learning based on their own experiences. ST's and SF's differ in that ST's look for facts, data and information in the details, and SF's look for ways to relate to new material on a personal or emotional level.

Intuitive learners

Intuitive Thinkers

If your highest number is in the NT quadrant, you are an intuitive thinker. The **iNtuitiveThinker** (NT) is characterized by logical thinking, perception of patterns, and a strong need to understand. An NT needs to mentally process new material alone before discussion, and must see the overall picture prior to processing details to build understanding. When you study alone, you should first look over new material to get the overall picture. Once you have the global concept, the pieces fit in naturally. You need to look for patterns in the information in order to help you remember, using mnemonics or other memory tricks as needed. Since NT's typically do not memorize well or easily unless they have an ST subdominant, you will need to understand the concepts in order to figure out the details that you may have difficulty remembering for tests. Many very bright NT's may have had trouble in school early on because they tend to dislike memorization and doing endless pages of drill on material they already know.

Since NT's see things in mental images in 3 and 4 dimensions, and usually think in pictures, they tend to gravitate towards and excel in engineering, sciences, architecture and design areas and research of many kinds. They can also excel in management or project management where an understanding of the inter-relationships of many parts of a complex whole must be managed simultaneously.

Intuitive Feelers

The **iNtuitive Feeler** (NF) is the creative learner. You may have a strong dislike for routine work, organization, and memorization, and be most at home in the abstract, doing creative things in fine arts or music. You are probably a daydreamer. You might even have considered dropping out of school out of frustration. Don't do it! The most famous NF's were Albert Einstein and Leonardo DaVinci. Even though many schools aren't designed to handle creative thinkers like you, that doesn't mean you aren't intelligent and capable. Some of the most brilliant inventors, artists, and designers are NF's, and many almost flunked out of school because they didn't fit the mold of what instructors expect.

You probably learn best through metaphor, building new learning on a structure of comparison with some other concept that you know well, no matter how far afield or how strange the connection might sound to other people. In fact, you might be happiest when other people say "that's really strange." But it's okay, because it works for you. You may work well in cooperative groups, but not necessarily, and should study with other classmates whenever possible if that works for you. Working with others turns off some NF's if they feel too much criticism from other students. When you are working on new material, you should look for another situation in which the same "rules" apply. For example, "How is an internal combustion engine like a rock band?" has meaning if you consider the heat, light, sound and other energies, and you can use that to build your understanding of internal combustion engines by comparing and contrasting the two ideas. The key is to harness your daydreams to productive thinking, using the dreaming to find metaphors that work and original ways to complete assignments.

The intuitive learners, NT and NF, work most effectively in the abstract, and need to begin new material with a global perspective, starting with an overview of the concept, and then filling in the details. This direction may be just the opposite of what most of your instructors use. If you are an NT or NF you will need to read or at least skim the chapter material prior to class in order to get an overall view of the whole concept before the instructor starts with the details. That way you can "build your framework" and see how the details fit into place when you hear them.

We hope that eventually teachers will use a wide variety of teaching strategies in the classroom, since each classroom is likely to include some students with dominances in each of the four quadrants. But actually, it is always the responsibility of you, the student, to attack your learning in whatever way helps you to learn the material and to earn your grades.

Chapter 3: Study Techniques by Cognitive Profile

You have probably been taught study techniques somewhere along the line by a parent, teacher, or guidance counselor. You may have been taught to choose a place away from distraction (no TV or radio), with a straight chair and clean desk, use good lighting, and have all your study needs, pens, pencils, papers, calculator, etc. ready before you begin. All that is probably okay, with the possible exception of the straight chair, the desk, bright lights, and maybe the absence of the radio or music, and you might be better off studying with a friend. You should, however, have all your necessary materials together before you start. What those words of wisdom don't consider is that we all are individuals, and we all have different needs. Some of us need comfort, and the straight chair would be a distraction when it got hard and uncomfortable. Like the three bears, some of us like it warm, and some need it cool. The "just right" temperature might be very different for you and your roommate. Some of us need music, but it should be instrumental, and should not be something you know the words to. Words, a song that is particularly meaningful to you or one with an intrusive beat will be a distractor. According to recent research (Rauscher, 1993, 1994; see Appendix B) Mozart is particularly good study music. Basically, whatever environment works for you is fine, provided you leave the TV off. If you need the background noise, try classical music. It doesn't distract as a TV program will, and some researchers say it induces the right kind of brain waves to assist learning. I especially suggest you don't use music with a driving beat as it is likely to be as distracting as words. Basically, anything that takes your attention off your primary reason for being there is a distraction.

But environmental issues only address the outside, the creature comforts. Being comfortable when you study might get you to study a little bit longer, but won't really help your grade on that exam. Once you are comfortable, we need to talk about what happens *behind* the eyeballs, because that's where learning happens. And that is different for each type of learner.

Before we begin the next section and discuss the quadrants one at a time, there is one more thing to mention about quadrants: although you have a dominant quadrant, you do have a complete profile with some area in each quadrant. You might find that you need to use different quadrant study techniques depending on which course material you are studying. Use your dominant quadrant first and most often. If you have a lot of material that you need to memorize, you might try ST techniques, or if you need to do some creative work, writing or designing, try NF techniques. Don't be afraid to work in a quadrant where you don't have a lot of strength. Ideally you'll build skills in those other quadrants and expand your capabilities, possibilities, and horizons. You should especially try to build skills in the intuitive areas, as those are necessary in the higher order thinking skills.

ST Strategies:

You are working in your Sensor Thinker (ST) quadrant when you memorize lists of information, such as names, dates, and places in history class, or items for a grocery list you need to get from the store, or the names of constellations on a clear night.

If your profile shows dominance in the ST quadrant- that is, if the ST quadrant has the greatest amount of area after you play dot-to-dot and draw in your quadrilateral profile- you probably are most comfortable memorizing what you will need for the test. If this is you, or if the material that you need to learn demands this treatment, there are certain things you can do to make it easier.

First, you will probably be most comfortable studying alone, in a place that is quiet and free from distractions. You also probably like good lighting and a conventional table and chair. If you want music on (you probably don't, but if you do) be sure that it is instrumental only. It also should not be anything with which you have strong memories or personal connections. So as much as you like it, don't play "our song" over and over. Many students find that Mozart's instrumental works are good study music. (Rauscher, 1993, 1994; see Appendix B)

Next, you need your study area to be well organized. Be sure you have everything you will need at hand and arranged in a way that suits you before you begin. If you will need a calculator, for example, get it out before you start. Your pencils should be sharpened, your light at the angle you like. In short, everything should be set the way you like before you begin.

As for the material itself, begin reading at the beginning of the chapter, taking care to assemble the details in order. Take notes while you are reading, either in outline form, or as numbered lists of facts. Every so often, perhaps every 5 to 10 minutes, or by sections in the text, stop and review what you have done so far. Don't let too much time go by before these little reviews. If you find you need to stop more often than 5 minutes, do so. If 10 or 15 minutes works for you, that's fine too. But don't go more than 15 minutes at a stretch unless you are one to hyperfocus. If there are problems to be worked, work a few of them each time you stop. If there are vocabulary words or terms, practice spelling the words, writing them out with the definitions a number of times until you know them well.

For difficult or complex material, you may find it helpful to make up flash cards of the important terms or concepts as you study, which you can carry with you for added review as you wait on line for meals or whatever odd moments. Break the material into manageable pieces, and work to get one piece down at a time, before you go on to the next piece. Remember, you are working to understand the material, not to just be able to regurgitate it.

If there are complex calculations to work:

1. Lay out a sample problem on evenly lined paper, preferably quadrille.

2. Identify the steps you go through to solve the problem, and what happens to each variable in the process.

3. Write out instructions for each step, not in terms of "plug and play" but in terms of the meaning of each variable.

4. Remember you are learning to solve the problems, not just plug numbers into an equation.

After you have done this, practice solving similar problems, and problems where different variables are given, where you must solve for different variables than in the sample problem. Look at word problems from the chapter, and practice setting up the problems according to your steps.

For the ST learner, practice makes perfect. Just be sure you practice each of the many different types of problems, not all the same type.

Some parts of the ST and SF study skills are quite similar because both ST's and SF's are concrete learners. The major differences lie in ST's doing best alone, and SF's doing best with company. If you are an ST and your best friend is an SF, you will need to be tactful in telling her that you study best alone, but after you have it down, the two of you can work together on practice problems and drills.

In Summary ST learners should:

1. Build from details to concepts

2. Organize material and use lists

3. Create stepwise problem set-ups (chunking)

4. Do repetitive drill

5. Memorize

SF Strategies:

The Sensor Feeler (SF) learner likes to talk things through, and wants to know how the material affects her life. How is it like something familiar and comfortable? You are functioning in your SF quadrant when you identify on a personal level with the material or subject. The SF learner generally learns best when working with another SF learner. When SF learners talk through the material, what happens is a specific cognitive process, where the SF is forming sentences about the content, using the vocabulary and saying something. Sometimes they will not be complete sentences, and frequently at first they won't make a lot of sense or be strictly correct, but the sentences must be made and must be spoken out loud. The correctness and sense will come later as understanding improves. I tell my SF students, if you study alone, talk to yourself. Talk to the walls. Talk to a mirror. Call you mother and tell her about it. She probably won't understand everything you are talking about, but she'll be glad to hear from you (and know that you are studying) anyway! When you begin new material, begin reading at the beginning of the chapter, taking care to assemble the details in order. Take notes while you are reading, either in outline form, or as numbered lists of facts. Every so often, perhaps every 5 to 10 minutes, or by sections in the text, stop and review what you have done so far. Talk about what you have read and explain the concepts. When you think you understand it thoroughly, write out a paragraph as if explaining it to a 5 to 7 year old child. (You don't have to have one of these children as a study assistant, an imaginary friend will do nicely!)

Don't let too much time go by before these little reviews. no more than 10 minutes is preferable. If there are problems to be worked, work a few of them each time you stop. If there are vocabulary words or terms, practice spelling the words, writing them out with the definitions a number of times until you know them well. Break the material into manageable pieces, and work to get one piece down at a time, before you go on to the next piece. Remember, you are working to understand the material, not to just be able to regurgitate it.

If there are complex calculations to work:

1. Lay out a sample problem on evenly lined paper, preferably quadrille.

2. Identify the steps you go through to solve the problem, and what happens to each variable in the process.

3. Write out instructions for each step, not in terms of "plug and play" but in terms of the meaning of each variable.

4. Remember you are learning to solve the problems, not just to plug numbers into an equation.

5. Practice solving similar problems, and problems where different variables are given, where you must solve for different variables than in the sample problem.

6. Look at word problems from the chapter, and practice setting up the problems according to your steps.

Some parts of the ST and SF study skills are quite similar because both ST's and SF's are concrete learners. The major differences lie in ST's dong best alone, and SF's doing best with company. If you are an SF and your roommate is an ST, you will need to accept her need for quiet, and not try to talk to her while she is studying. Find another SF to study with!

In summary SF learners should:

1. Study with a buddy

2. Talk it through

3. Personalize/personify content

4. Practice

NT Strategies:

The Intuitive Thinker (NT) learner spends much of the early educational experience frustrated. Often considered (usually inappropriately) "late bloomers" educationally, the NT needs to know the destination before planning the trip. Many NT's recall the common instructor's litany, "You have to build the foundation before you put the roof on the house!" The biggest educational obstacle for the NT is the conventional pedagogical wisdom of details first, building to concept, which is the exact opposite to what the NT needs to do.

The NT learner needs to see the overall concept first. Begin a new chapter by reading the abstract on the first page, then turn to the end of the chapter and read the summary to get an idea of what major ideas will be covered in the chapter, and how they all fit together. Also familiarize yourself with a vocabulary list, or list of equations. Then in the chapter, look over the illustrations, read the captions, and look at any charts and tables. Charts will often show you relationships between variables.

After you have a rough idea of the overall concepts, then go back to the beginning of the chapter and read it through. You may need to reread sections after you get to the end to help fit the pieces together, since most texts are not written with your "whole to part" learning patterns in mind. As you read, pay attention to where these details fit into the overall concepts you picked up from the summary.

It is helpful to take notes while you read. You will likely find your notes full of pictures and diagrams which show the patterns in the information. Lines and arrows are meaningful to the NT, and help to order the content into the 3-dimensional patterns of reality. When you encounter calculations, look at the variables and at what is happening to each variable in each step of the calculation. See in your mind's eye what is being calculated. Picture what is begin changed in the system.

NT's may or may not care to work in groups, but rather usually prefer to do it alone first, then to come into the group once they thoroughly understand what needs to be done and how to do it. NT's should therefore attend to their own behavior and try NOT to take over the group, which is a natural tendency, since they may have already determined how to accomplish the task. For this reason, NT's make good tutors in a peer group, explaining the physical and three dimensional concepts the ST's and SF's may have initial difficulty seeing.

In summary NT's should:

1. Start at the end

2. Get the overall pattern first, then fit details as they come.

3. Find logical patterns in problem solving

4. Emphasize understanding "Why?"

5. Do some of each problem type to show you understand.

6. Try new problems, and come at them from different directions.

NF Strategies:

The NF learner is the most unique, and the least understood in most academic environments. Global in perspective and poetic in nature, the NF needs to see the whole picture at the beginning to get an overall view of the idea.

Like the NT, the NF learner needs to see the overall concept first. Begin a new chapter by reading the abstract in the first pages, then turn to the end of the chapter and look for the summary. Read the summary to get an idea of what major ideas will be covered in the chapter. Look over the pictures, read the captions, and look at any charts and tables to get a feel for how the information fits together.

After you have a rough idea of the overall concepts, then go back to the beginning of the chapter and read it through. You may need to reread sections after you get to the end of each to help fit the pieces together, since most texts are not written with your "whole to part" learning patterns in mind. As you read, pay attention to where these details fit into the overall concepts you picked up from the summary.

As you read, take notes in whatever form is meaningful to you. Your notes might be mostly pictures and diagrams, but you do need to include the important vocabulary so you are able to communicate your understanding of the content to others. Draw pictures to show what is happening in each equation, visualize and plot data as it would change over time or with a change in one variable or condition.

When it is necessary for you to remember details, you can try this colorful technique. Write out the equation or diagram or structure you will have to recall on a fresh page. Then retrace the lines with another pen, then a pencil, then a variety of colored writing implements, colored pens, pencils, crayons, or markers. Each time, carefully trace the lines in the same order each time. Later, when you need to recall it, you should find you can look up into space, or close your eyes, "see" the image, and read it right off the image of the page in your mind. If you repeat the physical pattern of the retracing each time you may also have the kinetic memory and be able to write it out on scrap paper. Take great care if you do this that it is obvious that you did not bring the paper with the equations into the room, perhaps by writing them out on paper your instructor hands out for the test.

Find metaphors for the larger concepts to help you to build understanding for how things work. You probably find yourself saying things like "That's just like ... but this has A and that has B." These metaphors are beneficial constructs and very helpful to you, even though your study partners see them as diversions.

Eventually you will have to work through calculations. You may be able to "see" the 3- and 4-dimensional changes that are taking place as the equation describes what is happening. You may imagine yourself inside the changes taking place. Imagination is one of your strongest tools. Use your imagery and metaphorical reasoning in creative study sessions. Work towards developing a depth of understanding of the concepts. Middle or high school teachers and even an occasional college professor may be pleased to accept creative adaptations of routine assignments. Just clear your ideas first to ensure you get full credit.

In summary NF's should:

1. Start at the end

2. Get the overall picture first, then paint in details

3. Use metaphorical descriptions

4. Use mental images

5. Emphasize "what if?"

6. Be creative and try new things, no matter how strange.

7. Find something you already know that relates to the new material, and figure all the ways it's the same and all the ways it's different.

Chapter 4: Choosing a Major and a Career

College Major and profession

I have found, mostly by giving the inventory to people in a variety of professions and thousands of students in a variety of majors, that there are strong correlations between dominant quadrant in the cognitive profile and choice of a career path. It all makes sense when looking at what types of tasks people in each quadrant are good at.

Sensor Thinkers

ST's, Sensor Thinkers, are organized, methodical, and linear thinkers. They tend to major in areas that depend on memorization and attention to detail rather than logic and conceptual thinking unless they are also strong in the NT quadrant. ST's most often go into business or liberal arts, languages, secondary education especially in history, or some types of engineering. It is common for ST's to have done well in mathematics or sciences in high school by memorizing equations and plugging in numbers, but in higher level math and science courses may find themselves suddenly struggling. Bright ST's are well advised to develop skills in the NT region, such as visualizing a 3- or 4- dimensional figure (the fourth dimension is time) in a mental image and rotating it. If you can't do this, you won't do well at advanced level sciences and mathematics that require conceptual understanding and logical problem solving.

ST's may do also well in business, banking, law, pharmacy, or medicine. Medical school and law schools both require a great ability to memorize and retain large amounts of information. The major science in medicine is biology, which frequently can be aced through memorization with a modicum of NT logical analysis. There is a thought that the more area in a secondary NT quadrant, (an NT subdominant) the better diagnostician the doctor may be, more able to see the problem from several sides at once. The ST can do repetitive tasks, perhaps very complex repetitive tasks, without variation. Other professions include airline pilots, for the same reason. A pilot needs to land the plane the same way he's done it a thousand times, successfully, rather than get creative and try something new to vary the job. Similarly, if you were having open-heart surgery, you would not likely want the surgeon decide to get creative, and "try it from a different angle for a change." You need him to do it the same way he's done it before, successfully.

The danger for someone with too much area in the ST, to the exclusion of other quadrants, is a tendency to become obsessive. If you find you are obsessing, getting overly stressed, or being too demanding of yourself and others, try to draw back from the ST exclusion and do something creative on a regular basis. Take a music, dance, photography, creative writing, or drawing studio class. Get outside for a walk, with no destination or goal in mind. Do something outside of your regular patterns. ST's also tend to think that there are two ways to do things, their way and the wrong way. If you find yourself falling into this category, then get up and go play in the NF quadrant until you can back off a bit.

Sensor Feelers

SF's, Sensor Feelers, are people people, as the old song says. They go into professions and occupations that allow them to be in contact with people, usually in a helpful capacity. They are compassionate nurturers, and caregivers. Popular majors are nursing, physical therapy, and health care of any kind, elementary education, sociology, social work, and the like. Unless strong also in the ST or NT quadrants, they generally avoid mathematics and sciences. Other choices are marketing, sales, communications, and human resources. A balanced SF/ST may do well to consider pharmacy.

A particular note of importance is relevant here. I have found that nursing students are predominantly SF learners, and nursing professors are strongly ST. This disparity between the nature of the students and the expectations of the faculty can cause tension in the course of study. Nursing students must realize that preparation for their chosen, nurturing field requires a lot of individualized hard work, attention to details, and memorization and understanding of important information. Their patients' lives depend on their being able to carry out detailed tasks correctly every time, which is a decidedly ST skill.

If a person becomes too strongly SF to the exclusion to other quadrants, there is a tendency to worry excessively about others. If you find that you do this, you may be able to center yourself a bit by doing something that requires you to think through a problem, work a jigsaw, or solve a crossword puzzle. Organize your files or drawers and closets, put things in a logical, linear pattern.

Intuitive Thinkers

NT's, Intuitive Thinkers, are logical, need patterns and prefer a global perspective. They enjoy puzzles and challenges, and have a profound need to understand "why." They are problem solvers, and therefore often find themselves in engineering. They make excellent teachers, mechanics, or builders because they tend to see the whole picture at one time and can plan in multiple dimensions. An NT with experience in a particular business generally makes a superior manager, as NT's can see all sides of an issue or problem to direct the efforts to the most efficient and effective solution. Although not necessarily socially outgoing, they are natural leaders but do occasionally need to curb their tendency to take over and lead those who would rather not be led.

If the NT dominant has a strong ST subdominant, the directions to consider are industrial engineering, manufacturing, process control, quality control, electrical engineering, research and development with an emphasis on development. In addition to engineering fields, the NT might consider law enforcement, or forensic science. If the subdominant is NF, then the choices within engineering might have more to do with research, with greater demand for creativity in contrast to daily repetition of a manufacturing environment, which may feel too confining. NT's also make excellent secondary education teachers especially in mathematics and the sciences, computer gurus, research scientists of all types and architects.

NT's who are too far into patterns and logic to see their way out may begin to see patterns where none exist. Some may call it paranoia, but "just because we're paranoid doesn't mean they're not really out to get us!"

If you find that you are looking behind you at every corner, or otherwise feeling cornered at work, take on some activities in the opposite corner of your profile, the SF. Hang out with some friends, socialize, take the weekend off for a change and go on a road trip. At least organize a bunch of friends to go out to lunch, or for some activity you all enjoy.

Intuitive Feelers

NF's, Intuitive Feelers, are creative, above everything else. They think in pictures or mental images, often experiencing several senses as they build a memory of new material. An NF cook may taste the finished dish in his imagination as the herbs and seasonings are added, allowing him to develop a new recipe. My recent research has shown a strong correlation between NF dominance and diagnosis of Attention Deficit Disorder with or without Hyperactivity, or ADD/ADHD. The list of behaviors that are used as diagnostic for ADD is almost identical to the list for NF dominant learners. Most of the young people identified by their schools and pediatricians as ADD may not be ADD at all, but NF learners who are out-of-step with the normal classroom routine. The telling difference is whether the NF dominant learner is able to concentrate for extended periods of time on things he is interested in, rather than on assigned tasks.

They are typically artists: painters, writers, poets, musicians, photographers, teachers of English, fine arts, drama and music, designers, decorators, window dressers, or architects. If the NT is a strong subdominant they are often architects, research or design engineers, inventors, or environmental engineers. If the SF is the subdominant, they may feel most at home in sociology and social work, psychology, marketing, or advertising. They need a profession that encourages the creativity they exude and that is not clock and calendar driven. They are passionately involved in causes, and are often devoted to saving the whales, the rainforest, and humanity from itself in alternate weeks. They tend to make emotional decisions rather than logical ones.

If you find yourself, the NF, too far off on a tangent or lacking touch with reality, tackling a linear task may get you back on track. Make a list of what you need to do, prioritize the tasks, and then make a specific plan to accomplish those things on the list by a specific date or time. When you have done a predetermined number of tasks, reward yourself with some time or activity you enjoy.

Chapter 5: Coping with Teacher Types

You will meet all kinds of instructors in your educational experience. Just for fun, and perhaps a little more enlightenment, I want to try to point out a few of the more common types of instructors and the best way to deal with them.

Keep in mind that instructors have types and styles just as students do. People don't usually outgrow their types. If anything we become stronger in one quadrant and less strong in the others as we find our niche in life and concentrate on one subject area. You will probably come across each of these or combinations of them as you go through college.

THE DRONER- Drones on in a monotone lecture, using the same lecture notes and audio visual materials year after year. Hardly knows students are in the room. Looks at the back wall as he speaks. Don't even try to ask questions in the Droner's lecture. Instead, visit the instructor during his office hours. Have a reason for your visit, such as further clarification on something in the text or an assignment. Introduce yourself, being as engaging as possible. Show interest in his research or hobbies if any are apparent, then proceed to your agenda.

DR. FOREIGNER: Writes long equations or explanations, working feverishly on the board. Has a strong accent, and if you can't understand him, it's your problem, not his. Usually lectures facing the blackboard. Seems to enjoy erasing the board to start filling it up again before you have any hope of getting it all written down in your notes. This instructor has taught at every university I personally have attended, and from what I hear, every other university in the country too. You have little chance of avoiding him or her entirely. Your only hope is to read the chapters ahead of time so you are familiar with the vocabulary and know what to expect, and try to follow his line of thought. Get down as much as you can, but concentrate on the concepts, and understanding what the equations or visuals are describing. Reading and preparing thoroughly before class will make the difference between life and death with a non-native speaker's course. If you know what words to expect, you will be able to understand more of what he is saying. If you are familiar with the material, you will be able to take your notes with less time bobbing your head up and down to look at the board. (I have no direct experience here, but I suspect Americans may do the same things when teaching in non-English speaking countries.)

THE LEAPER: Leaps from one thought to another with hardly a passing glance to see if anyone is following. Lots of energy, but confusion reigns. Genuinely likes students, but wants to share so much his students get lost in the clutter. Your best hope is to read the summary material in the text before he starts a new chapter. Then you'll know where he's going, and how it all fits together. If you are an ST, you'll want to make a brief topical outline of the chapter before class and leave lots of space in your outline to fill in other notes. Then you can skip around in your outline making notes as he leaps, and your notes will stay in linear order.

THE ARTIST: Runs to and from the board, drawing stick figures, lines and circles to explain the concept. Disconnected fragments of sentences tumble over one another, confusing you and your classmates. The key here is to connect with the concept. Make a mental image to match his sketches, taking down his diagrams in your notes. Read the chapter summary ahead of class so you have a basis for comprehending what he's rambling about, and then re-read the text material with your notes and the diagrams in front of you, filling in and labeling the diagrams as you build understanding.

THE STORYTELLER: Everything has a personal connection and brings back memories for this instructor, or gives him an excuse to tell a story that may or may not have connection to the lecture topic. Good for SF's, but may be hard to follow for the rest. Instead of hearing the story for the sake of the story itself, listen with the content under study in mind, so you hear the connections. The instructor is probably telling the story to illustrate something about the concept. Don't get lost in the events or people in the story and lose the concept the story is intended to exemplify.

IL DUCE: There are two ways to do things, his way or the wrong way. He's the professor, and don't you forget it. Don't try to argue, you'll only lose. You'll likely find there are definite right and wrong answers to every question. You will need to accept that for this course what he says is right and give him the answers he wants, even if you have a real basis to question the validity. You could try to engage him in a debate, but be ready to back off if this appears fruitless or accept the risk of an F. If you can't deal with his way of teaching the course, you might be better off dropping the course and taking it another time with someone else. Check first that it is offered with someone else. Often Il Duce will be the only one who teaches his course, as in his eye, no one else would do it right.

There are many others, of course, and most of us are a mixture of several types. Know your own profile, and try to adapt your way of learning to your instructors' way of teaching. Reading the material ahead of class is always a good idea, and sometimes can make the difference between an A and a C or worse. And remember also, it's not a bad idea to drop a course and take a lighter load for a semester if you are really in trouble. The phrase that pays is "Protect your GPA!" Prospective employers or graduate schools may look harder at your GPA, than at how long it took you to get through school, especially if you have family or other responsibilities at the same time.

CHAPTER 6: THE ADD/NF CONNECTION

Let me start by stating unequivocally that there is a very real medical condition that is known as ADD, attention deficit disorder. It involves a malfunction in the blood supply in the brain such that when a particular type of mental task is attempted, a part of the brain that needs more blood supply actually gets less. The harder you try, the more difficult it gets. The only way ADD can be definitively diagnosed is when an MRI (Magnetic Resonance Imagery) or EEG (electroencephalogram) is taken under mental stress. For an MRI you lay down with your head in a huge machine that takes X-ray-like pictures of the blood flow in the brain while you are given problems to do or puzzles to figure out in your head. For an EEG you have wires attached to your head (with adhesive pads) while the technician takes the readings of electrical activity in your brain.

The tests are very expensive, and although few people have done it outside a research setting, estimates have been made that perhaps 0.5% of the population suffers from real ADD. Some of those people also have hyperactivity, which then is called ADHD or attention deficit hyperactivity disorder. (Some psychiatrists call it just ADHD of 3 types, inattentive, hyperactive, or combination.)

We currently have about 15% of our school children in the United States diagnosed with ADD or ADHD. The overwhelming majority of people diagnosed with ADD or ADHD are diagnosed based on a list of behaviors, which are rated by their parents or teachers. The ratings are totally subjective, and there is no way to separate out how much is based on different expectations of the parents or teachers. In many cases, especially where ADHD is the conclusion, the child is prescribed Ritalin or another amphetamine to slow down the brain functioning and allow the child to fit in better in the classroom setting and be more manageable. Of course these drugs can change sleep patterns, reduce the amount of REM sleep (the good kind of sleep) and can lead to serious psychotic disorders.[1]

Some years ago I noticed that a large number of my students who were diagnosed with ADD or ADHD were coincidently NF dominant learners. A grant and a research study later and the results were confirmed. There is a statistically significant correlation (at better than 99% confidence level) of NF dominance with a diagnosis of ADD or ADHD. That does NOT mean that NF dominant learners are ADD. It means that NF dominant learners get diagnosed as ADD. Let's take a look at why.

Parents and teachers are asked to rate the child on a list of behavior characteristics. The list includes most if not all of the behaviors and thinking patterns that are part of the NF learner quadrant. If you are an NF dominant learner, you will answer many of the questions for ADD/ADHD diagnostic tests in such a way that you WILL be diag-

---

[1] Kaufman, David Myland, Clinical Neurology for Psychiatrists, 5th Ed., W.B. Saunders Co., NY, 2001

nosed as ADD or ADHD. In fact, I'm an NT/NF learner. In taking several of the tests online, I rate so highly on the lists that I get diagnosed as ADHD on every test I have taken.

The criteria for ADHD in the DSM-IV (the official book for psychiatrists and psychologists that describes mental health problems) include inattentiveness, impulsiveness, and hyperactivity.[2] NF learners may find it hard to pay attention to long winded lectures since their natural way of thinking jumps from one chunk of concept to another quickly, in what appears to be a randomness defined only by connections that the learner sees on the fly. NF's need the concept first in order to keep track on a large number of details. And the DSM-VI says if you can't keep track of details, you score another point towards a diagnosis of ADHD.

NF's also tend to be brighter than the average bear, or brighter than average bears tend to be NF's, the correlation isn't clear, and so the usual s...l...o...w... presentation of material in the classroom may cause him or her to drift off in a daydream. Oh, and daydreaming is another symptom of ADHD of the inattentive variety.

Many of our schools (elementary and middle schools especially, but even many college professors) aren't very good at dealing with intuitive learners, and can be even worse at serving the needs of the espe-

---

[2] AMERICAN ACADEMY OF PEDIATRICS, Committee on Quality Improvement, Subcommittee on Attention-Deficit/Hyperactivity Disorder, *Clinical Practice Guideline: Diagnosis and Evaluation of the Child with Attention-Deficit/Hyperactivity Disorder*, PEDIATRICS Vol. 105 No. 5 May 2000

cially bright students.

So if your profile shows that you are an NF learner or if you have a significant NF area in your profile, and if you have been diagnosed as having ADD or ADHD, think twice about it. Were you bored in school and daydreamed a lot? Did you get sidetracked easily by more interesting thoughts when you were supposed to be doing something that didn't interest you? Did repetition do you in? Did you not complete assignments that seemed either overwhelming to meaningless tedium? Are the reasons why you were diagnosed as ADHD included in the chapter on NF characteristics?

Try the study strategies that you have read for NF and see how much help they can be. Other students have seen huge benefits, such as going from barely passing in high school and feeling that there's something wrong with them, to straight A's in college and a 4.0 GPA with these strategies and a good understanding of what is really going on. The problem isn't you. It's the system. But the good news is that it's within YOUR power to fix it for yourself.

Are you taking Ritalin or some other amphetamine? You need to ask your doctor before you stop taking it, as this book cannot be considered medical advice. But you should ask your medical practitioner to consider your learning type in his or her evaluation.

The important thing is that NF learners have a difference in how they learn from the ST's, not a disability. If you have difficulty getting organized, let some helpful ST get you organized. ST's love to organize things. Let them! Once you have things in categories, it's a lot easier to keep them that way. You, in turn, can help the ST think through a problem or solve a puzzle.

A major problem we have in the educational world is that the language used when educators talk about ADD is "problem" language. They talk about typical NF behaviors as "no problem, mild, moderate, severe or profound." In my language they are different from ST, but not a problem. Since you are taking responsibility for your own learning, you need to realize that you may learn differently, but that's not a disability.

Appendix A - Glossary

Abstract:

Something that can be seen mentally, an idea, not a concrete object. Seeing something in your mind instead of as an object in front of you.

Analyze:

To study or examine something to see how it works or how it is put together.

Big Picture:

The whole idea at once, rather than individual details.

Concept:

A broad, general idea

Concrete:

Something that can be held, picked up, seen or touched.

Create:

To use imagination to make something new

Daydream:

To wander off in your mind away from what is going on around you.

Details:

All the individual little parts that make up a concept, idea, or thing.

Discover:

To find something new by searching for it.

Discuss:

Talking about something

Do for Others:

To give time, effort or money to others as first priority.

Doing:

Physically acting to make something or to make something happen.

Emotion , Emotional, Emotions:

Feeling

Enjoy doing:

The pleasure of making something or making something happen.

Experiment:

Trying something new to see what happens, when you don't already know what to expect.

Facts:

Things that are known to be true.

Feelings:
>Emotions

Get It Done:
>Finishing the task at hand.

Group activities:
>Doing things together with others.

Ideals, Idealistic:
>Living by philosophy, pursuit of perfection, or certain principles and before considering practical considerations.

Imagine, Imagination:
>Forming ideas and images in the mind, especially of things you've never seen.

Information:
>Facts, data, details.

Insights:
>Seeing a situation or person clearly and intuitively

Invent:
>To make something new, especially a functioning object, that has not been made before.

Join a group:
>Become a functioning member of a group, and working toward a group goal.

Law, Laws:
>Rules that are made by authority, that one can be punished for breaking

Lead a group:
>Function as the leader of a group, bearing responsibility for the group effort at the task assigned.

Listen to Music:
>Experience music as a rhythmic, pleasant relaxation. (Not intended as a social outing.)

Logic, Logical:
>Analysis of facts, data and ideas to reach a conclusion or decision.

Loyalty, Loyalties:
>Devotion, duty or attachment to a person or group.

Mental picture:
>An image in the mind, especially of an abstract idea.

Metaphor:
>A word or image of one thing used to represent another idea or concept, generally of two quite unrelated concepts.

Overview:
> A short description of a large concept or lengthy or complex lesson.

People:
> Social or professional interaction with others

Personal:
> About or within yourself

Plan it out, Plan it well:
> Getting the whole plan together before starting any physical action.

Possibilities:
> What can happen.

Practical:
> Considering actual facts, and real information

Practice:
> Repetition and drill.

Principles:
> Your personal ideals, rules.

Protect yourself:
> Protect yourself and your loved ones from harm as first priority.

Read A Book:
> Read a book for enjoyment.

Realistic:
> Practical consideration.

Relationships:
> Interpersonal, social interaction

Roles:
> Behaving in keeping with who we are in relationship to others.

Save the Earth:
> Protecting the environment and endangered species as first priority.

Sensible:
> Sound judgment and reason

Share, Sharing:
> Allowing someone else to use something of yours.

Solve a Puzzle, Work a Puzzle:
> Using mental skills to work a puzzle, as a crossword, logic puzzle, video game depending on puzzle solving (not fighting), or other game.

Specifics:
> Details

Strategy:
> Plan for accomplishing a task

Talking:

Being the speaker, or part of a discussion group.

Tell a story:

Relating a tale, either original or from a published story book, especially to an audience as entertainment.

Tell the Truth:

Honesty always and in everything, as a first priority.

Theoretical:

An idea, based on logical analyses and possibilities.

Think, Think about:

Reason, using facts at hand and analyses.

Touch, hold:

Being able to pick up an object, turn it over and around to examine it and explore it.

Trial & error:

Trying a wide variety of things to see what might work.

Understand:

To know or be able to explain why something is or works the way it does.

Visualize:

Be able to see a mental image of a thing, action or concept.

Appendix B

Mozart as Study Music:

Rauscher, F. H., Shaw, G. L., & Ky, K. N. (1993). Music and spatial task performance. Nature, 365, 611.

Rauscher, F. H., Shaw, G. L., Levine, L. J., & Ky, K. N. (1994, August). Music and spatial task performance: A causal relationship. Paper presented at the meeting of the American Psychological Association, Los Angeles.

1993 "MOZART STUDY"
Subjects

Thirty-six undergraduate students participated in 3 listening conditions.

Phase 1

All subjects listened to 10 minutes of each of the following listening conditions:

> 1. Mozart Sonata K. 448,
>
> 2. relaxation instructions, and
>
> 3. silence.

The listening conditions were presented in counterbalanced order.

Phase 2

Subjects were tested for spatial-temporal and spatial recognition skills after each listening condition.

Results

Subjects' spatial-temporal scores were significantly higher after listening to the Mozart Sonata. Spatial-temporal scores following the relaxation instructions or silence did not improve. Spatial recognition scores did not change following any listening condition. Note: The enhancement of spatial-temporal skills lasted only 10 minutes. One cannot conclude from this study that casually listening to music will induce long-term improvement of spatial skills.

References and Bibliography

The listed materials support the author's work and development of the Cognitive Profile Model, and the research and opinions expressed in this little book.

Krause, L. B. The Cognitive Profile Model Of Learning Styles: Differences In Student Achievement In General Chemistry. Journal of College Science Teaching. Vol. XXVIII, no. 1, Sept/Oct 1998

Web site: Cognitive Profile Learning Styles Model, http://www.cognitveprofile.com

The Cognitive Profile Model Of Learning Styles: Clemson Kappan, vol. 14, no. 1, Winter 1997.

An Investigation of Learning Styles in General Chemistry Students. Doctoral dissertation, Clemson University, August 1996.

Aldridge, B. G. (1992). Project on scope, sequence, and coordination: a new synthesis for improving science education. *Scope, sequence, and coordination of secondary school science, vol. 2. relevant research.* Washington, D.C.: National Science Teachers Association

American Psychiatric Association: Diagnostic and Statistical Manual of Mental Disorders, Fourth Edition, Text Revision. Washington, D.C. American Psychiatric Assoc., 2000

Anderson, J. R., (1995). Cognitive psychology and its implications. fourth edition. New York: Freeman

Andrews, M. H. & Andrews, L. (1979). First-year chemistry grades and SAT math scores. Journal of Chemical Education 56,4 American Chemical Society

Armstrong, T. (1994). *Multiple intelligences in the classroom.* Alexandria, VA: ASCD

Barger, R. R. (1984). Psychological type and the matching of cognitive styles. *Theory Into Practice* 23,1

Barnett, J. (1994). *Learning styles.* South Carolina Department of Education

Bigge, M. L. (1976). *Learning theory for teachers,* third edition. New York: Harper & Row

Briggs, K., & Myers, I. (1991). *Myers-Briggs type indicator" Form G booklet.* Palo Alto: Consulting Psychologists Press.

Campbell, J. ed. (1976). *The portable Jung.* New York, Penguin

Columbia College of Education, Missouri University, (1986). *Learning styles* (ERIC Document Reproduction Service No. ED 323 249)

DeMarrias, K., & LeCompte, M. (1995). *The way schools work, second edition.* White Plains, NY: Longman

Druckman, D. & Bjork, R. A. (1994). *Learning, remembering, believing: Enhancing human performance.* Washington, D.C.: National Academy Press

Dunn, R. & Dunn, K. (1978). *Teaching students through their individual learning styles.* Reston: Reston

The following two volumes are a compendium of reprints of articles published over a number of years in a wide variety of journals and reports. All are either written by or about the work of Rita and Kenneth Dunn and/or their students. Because of the large number of individual papers and the similarity of content, they are referenced here as compendia. These two volumes are available for purchase from Dr.'s Rita and Kenneth Dunn.

Dunn, R. and Dunn, K. (1994). *A Review of articles and books, part 1.* New York: St. John's University

Dunn, R. and Dunn, K. (1994). *A Review of articles and books, part 2.* New York: St. John's University

Dunn, R., Gemake, J. , Jalali, F. , Zenhausern, R., (1990). Cross-cultural differences in learning styles of elementary age students from four ethnic backgrounds. *Journal of multicultural counseling and development.* 18 (Apr.) (Included in Dunn (1994) part 2, above.)

Dunn, R., Beaudry, J., Klavas, A., (1989). Survey of research on learning styles. *Educational Leadership,* 46, 6 (Included in Dunn (1994) part 2, above.)

Dunn, R. and Dunn, K. (1993). *Annotated bibliography.* New York: St. John's University

Dunn, R. Personal conversation, 10.21.94, at teachers' workshop on learning styles.

Ehrman, M. (1990). Psychological factors and distance education. *American Journal of Distance Education,* 4,1

Fordham, F. (1991). *An introduction to Jung's psychology.* New York: Penguin

Gallagher, A. & De Lisi, R. (1994). Gender differences in scholastic aptitude test—Mathematics problem solving among high ability high school students. *Journal of Educational Psychology* 86, 2.

Gardner, H. (1993). *Multiple intelligences: the theory in practice* New York: Basic Books.

Ginsburg, H. P. & Opper, S. (1988). *Piaget's theory of intellectual development.* Englewood Cliffs, NJ: Prentice Hall

Hanson, J. R., (1991). Square pegs: Learning styles of at-risk students. *Music Educators' Journal* 78, 3 (Nov)

Hass, G., & Parkay, F. (1993). *Curriculum planning: a new approach.* sixth edition. Boston: Allyn and Bacon

Herrnstein, R. & Murray, C. (1994). *The Bell Curve: Intelligence and class structure in American life.* New York: Free Press.

Howard, D. H. (1992). *Use of the Myers-Briggs Type Indicator (MBTI) with physicians: a review of the literature.* (ERIC Document Reproduction Service No. ED 344 150)

Jung, C. G. (1970). *Analytical psychology, its theory and practice, the Tavistock lectures.* New York: Vintage Books. (original work published in 1936)

Jung, C. G. (1990). *Psychological Types.* ninth printing, revision by Hull, R. F. C. of the translation by Baynes, H. G. Princeton: Princeton

Jung, C. G. (1964). *Man and his symbols.* New York, Dell.

Jung, C. G. (1989). *Memories, dreams and reflections.* New York, Vintage.

Klein, S. B. (1987). *Learning: Principles and applications.* New York: McGraw-Hill

Kliebard, H. (1992). *The struggle for the American curriculum* 1893-1958. New York: Routledge

Melear, C. & Richardson, S. (1994). Considering relational learning style of African American students in science instruction. *The Journal of College Science Teaching,* Vol. 24, Mar/Apr.

Myers, I. (1962). *Introduction to type.* Gainesville, FL: Center for Application of Psychological Type

Myers, I. & McCaulley, M. (1958). *Manual: A guide to the development and use of the Myers-Briggs type indicator.* Palo Alto, CA: Consulting Psychologists press

Ozsogomonyan, A. & Loftus, D. (1979). Predictors of general chemistry grades. *Journal of Chemical Education.* 56, 3 American Chemical Society.

Pascal, E. (1992). *Jung to live by.* New York: Warner.

Rauscher, F. H., Shaw, G. L., & Ky, K. N. (1993). Music and spatial task performance. Nature, 365, 611.

Rauscher, F. H., Shaw, G. L., Levine, L. J., & Ky, K. N. (1994, August). Music and spatial task performance: A causal relationship. Paper presented at the meeting of the American Psychological Association, Los Angeles.

Rippa, S. (1992). *Education in a free society: an American history.* seventh edition. New York: Longman

Rule, D. L. & Grippen, P. G. (1988). *A critical comparison of learning style instruments frequently used with adult learners.* (ERIC Document Reproduction Service No. ED 305 387)

Russell, P. (1979). *The Brain Book.* New York: Plume

Skinner, B. F. (1953). *Science and human behavior.* New York: Free Press

Silver, H. & Hansen, J.R. (1980). *Teacher self- assessment manual.*
Moorestown, NJ: Hanson Silver Strong & Assoc. inc.

Silver, H. & Hansen, J.R. (1981). *Teaching styles and strategies.*
Moorestown, NJ: Hanson Silver Strong & Assoc. inc.

Silver, H. (1986). *Task Rotation Strategy.* Moorestown, NJ: Hanson Silver
Strong & Assoc. inc.

Smedley, L.C. (1987). *Chemists as learners.* Journal of Chemical
Education. 64 American Chemical Society

Spring, J. (1994). *The American school 1642-1993* third edition. New
York: McGraw-Hill

Strong, R.W., Hansen, J.R. & Silver, H. (1980). *Questioning styles and
strategies: Procedures for increasing the depth of student thinking.*
Princeton Junction, NJ: Hanson Silver Strong & Assoc. inc.

Sternberg, R. J. (1992). *Wisdom: its nature, origin and development.* New
York: Columbia University Press

Williams, L. V., (1983). *Teaching for the two-sided mind: A guide to right
brain / left brain education.* New York: Simon & Schuster.

Woods, D. R., (1993). Models for learning and how they're connected-
relating Bloom, Jung, and Perry. *Journal of College Science Teaching*
(Feb.)

Yale University Institute of social and policy studies. (1988). *Report on the
New York State Board of Regents' panel on learning styles* (ERIC
Document Reproduction Service No. ED 348 407)